AF230752

K-2 Math
Volume 4

© 2013 OnBoard Academics, Inc
Newburyport, MA 01950
800-596-3175
www.onboardacademics.com

ISBN: 978-1-939796-75-2

ALL RIGHTS RESERVED. This book contains material protected under International and Federal Copyright Laws and Treaties. Any unauthorized reprint or use of this material is prohibited. No part of this book may be reproduced or transmitted in any form or by any means, electronic or mechanical, including photocopying, recording, or by any information storage and retrieval system without express written permission from the author / publisher. The author grants teacher the right to print copies for their students. This is limited to students that the teacher teachers directly. This permission to print is strictly limited and under no circumstances can copies may be made for use by other teachers, parents or persons who are not students of the book's owner.

Table of Contents

Units of Measure

Key Vocabulary

weight

ounce

gram

lighter

mass

pound

kilogram

heavier

Ounces and Pounds

<table>
<tr><td>Abbreviations</td></tr>
<tr><td>ounces</td><td>oz</td></tr>
<tr><td>pounds</td><td>lb</td></tr>
</table>

Ounces are used to measure very light things.

Pounds are used to measure heavier things.

____

How many ounces are in a pound? ____________

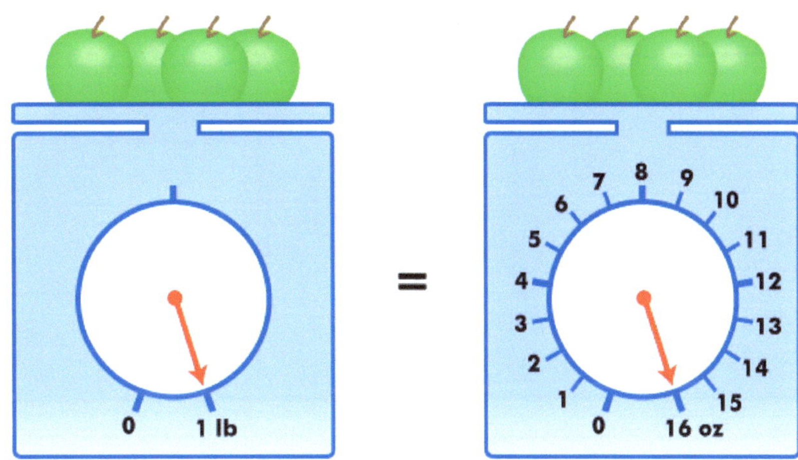

Where do these object fit on the number line.
Connect the red dots.

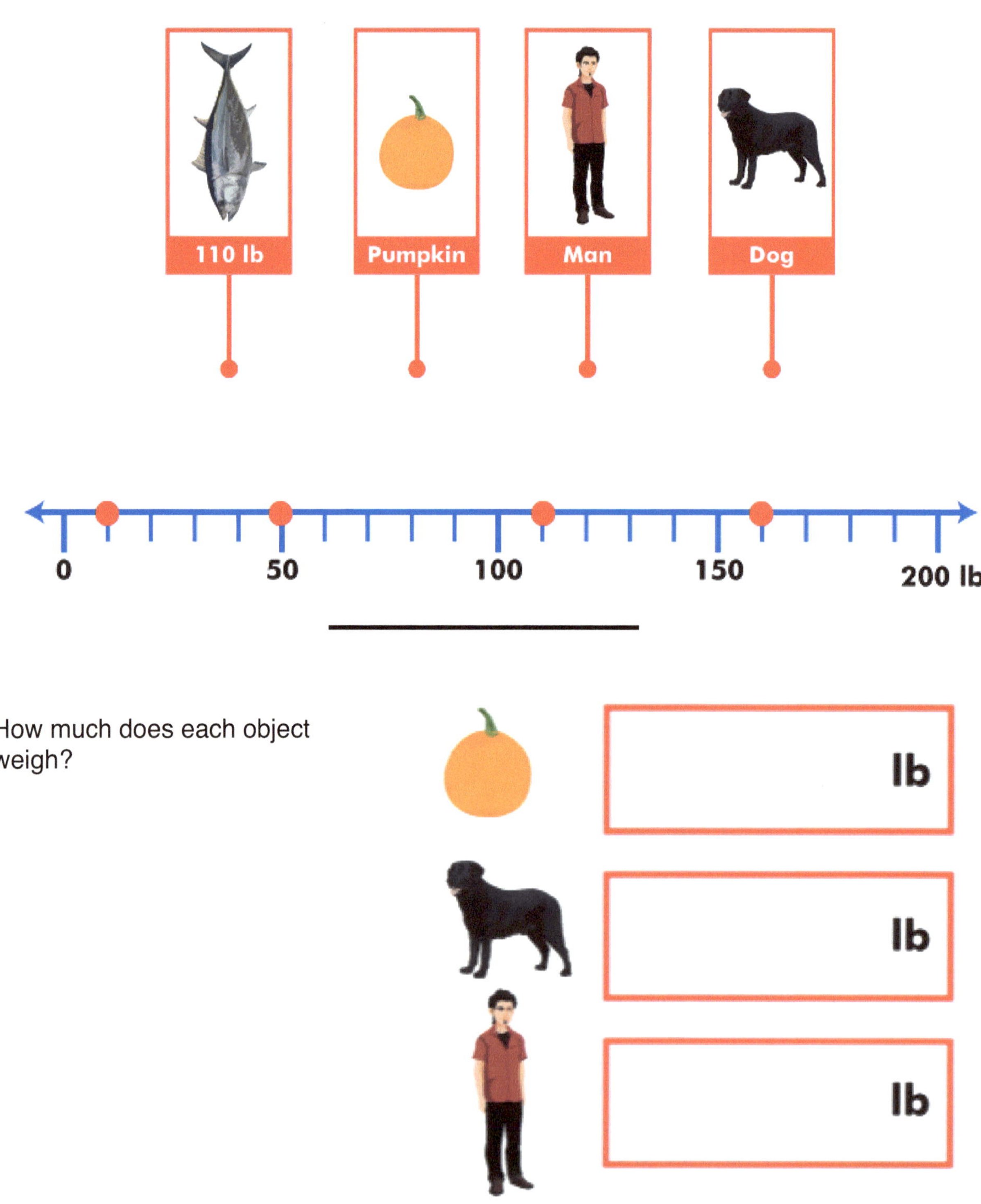

How much does each object weigh?

lb

lb

lb

Which piece of fruit is heavier?______________________________

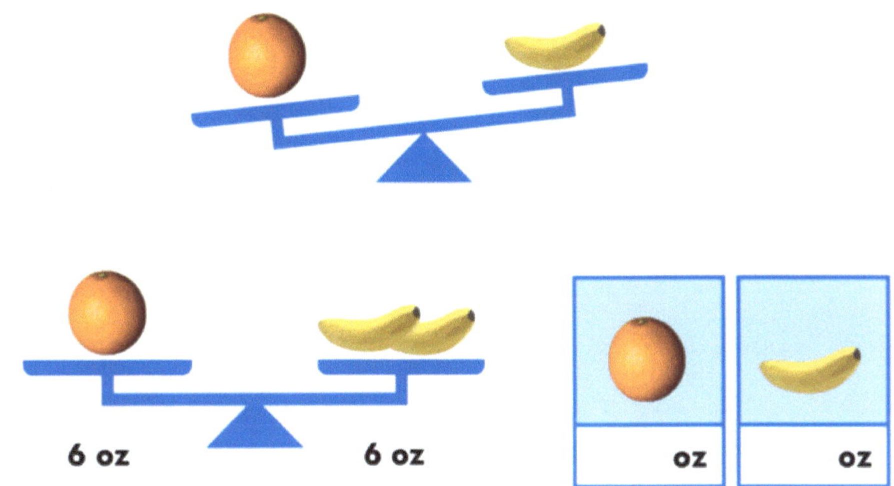

How much does each piece of fruit weigh?

____

Which piece of fruit is lighter?______________________________

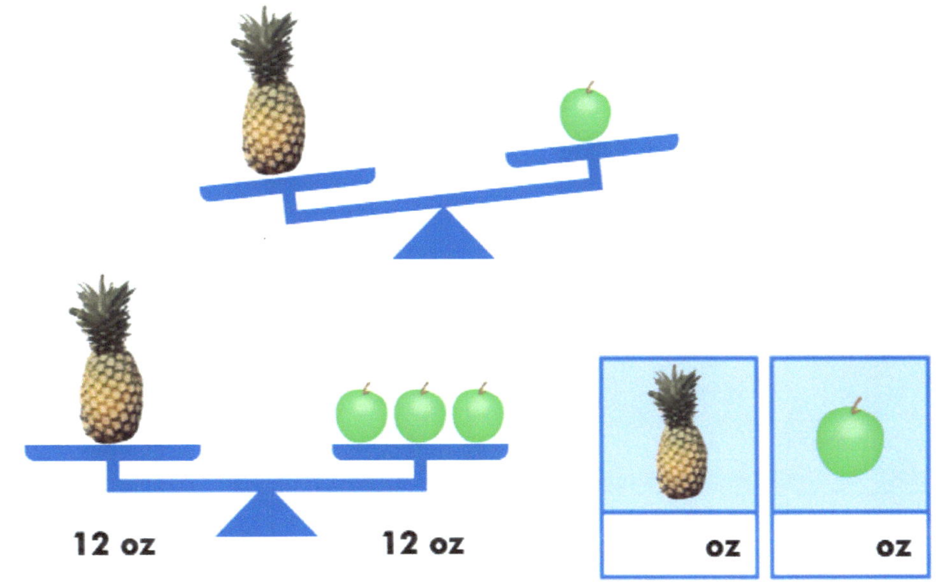

How much does each piece of fruit weigh?

Grams & Kilograms

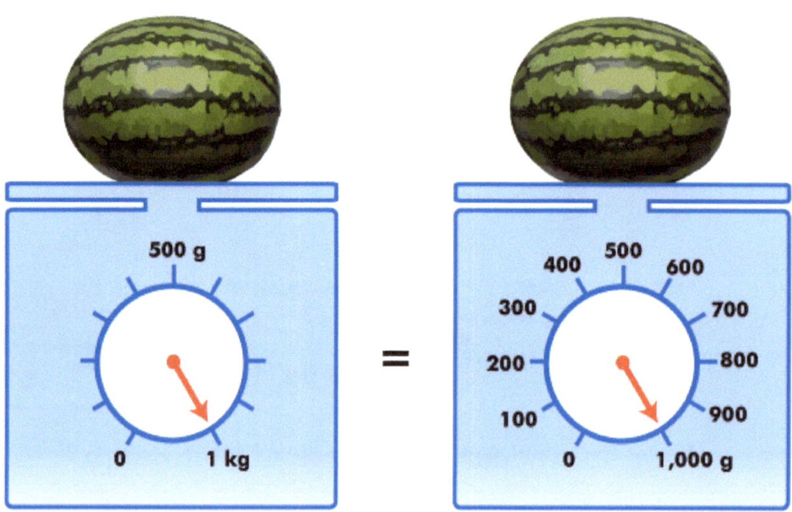

Do you know how many grams there are in a kilogram?

____

If there are 80 blueberries on this scale, what is the mass of a single blueberry and a single apple?

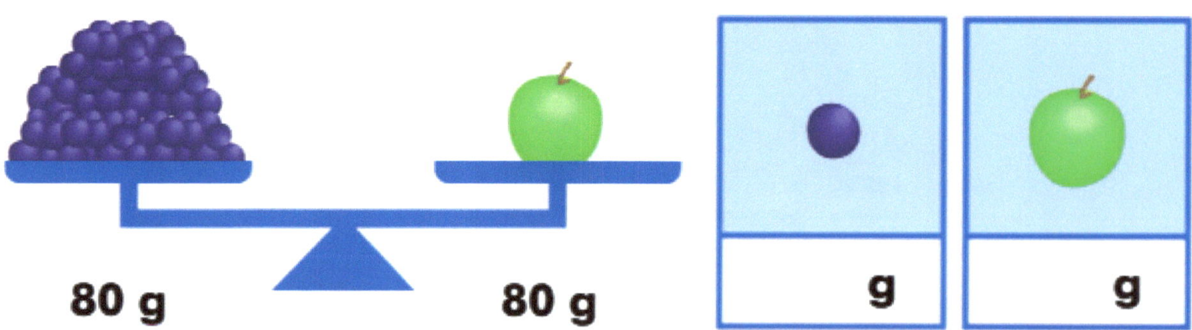

Units of Measure Quiz

Circle of fill in the correct answer.

1 **True or false? 400 g = 4 kg**

2 **Choose the best estimate for the weight of a cat.**

A **9 lb**

B **26 lb**

C **60 lb**

D **160 lb**

3 **2 lb = _____ oz?**

4 **5 kg = ___ g**

Time

Key Vocabulary

o'clock

minute hand

hour hand

colon

one-half

half past

The big hand and the little hand

What color is the minute hand? ___________

What color is the hour hand? ___________

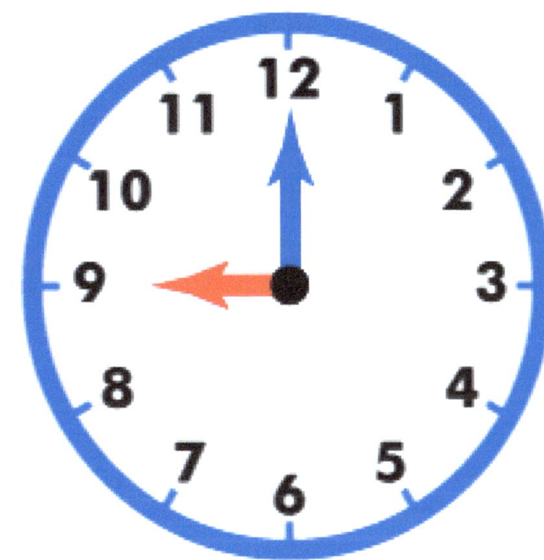

the time is nine o'clock

9:00

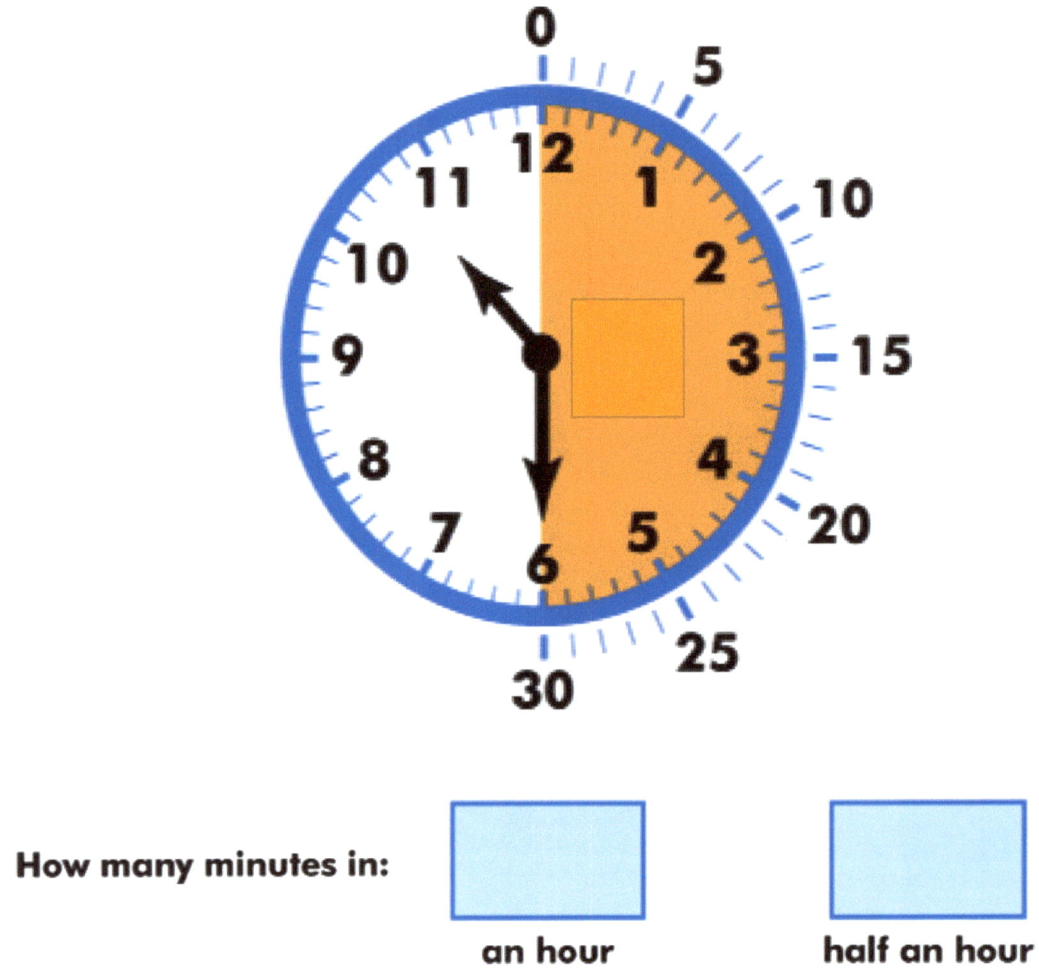

How many minutes in:

an hour   half an hour

Reading and Writing the time

Write the time in the boxes provided.

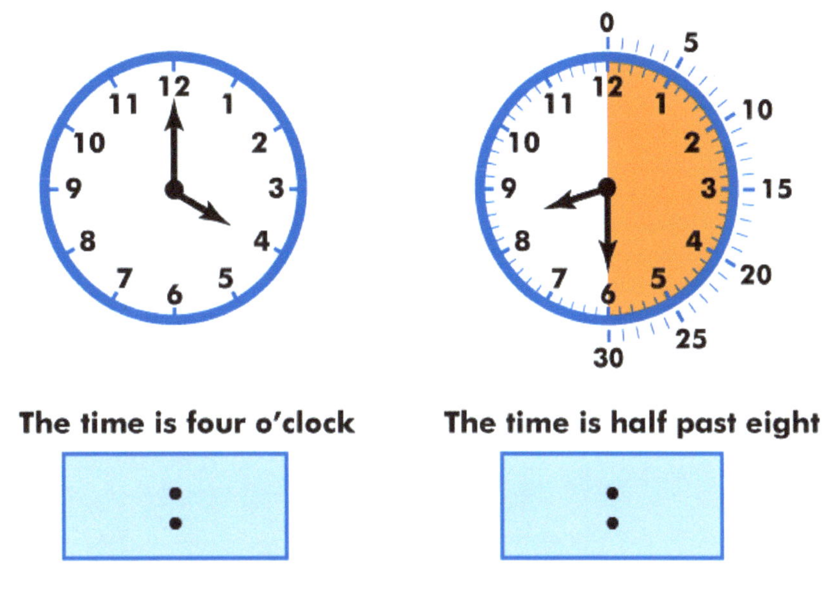

The time is four o'clock

The time is half past eight

What time is it?

Complete the chart below.

Set the clocks to match the times for Owen's favorite shows.
Draw in the minute and the hour hands.

Fill in the time and make a check in the am or pm

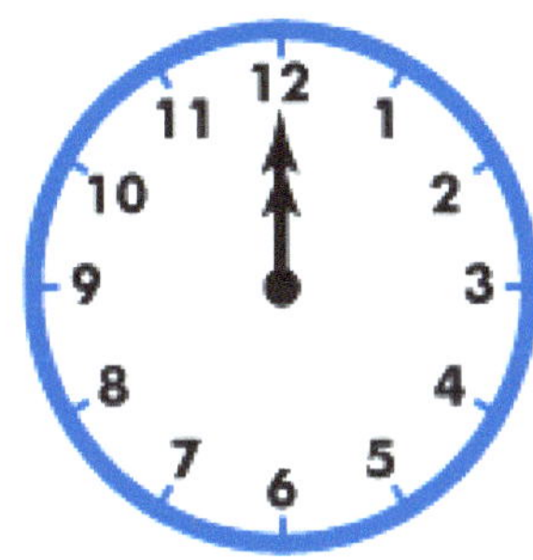

AM starts at 12 o'clock at night

PM starts at 12 o'clock during the day

"Hurry, or you'll be late for school!"

	:	AM	
		PM	

"We have one more lesson before lunch."

	:	AM	
		PM	

"It's time for dismissal."

	:	AM	
		PM	

Name____

Time Quiz

Circle or fill in the correct answers.

1 **True or false? There are 12 minutes in half an hour.**

2 **What is the time?**

A 5:30

B 6:00

C 7:30

D 6:30

3 **How many minutes is it until 7:00?**

4 **What number is the minute hand pointing at when it is 9:00 PM?**

Days, Weeks, Months, Years

Key Vocabulary

Days of the week

Months of the year

Can you order and count the days of the week?
Write the day in the proper box and add its day number for the week.

1	Sunday
3	Tuesday
5	Thursday

Friday

Monday

Saturday

Wednesday

B	A	S	U	N	D	A	Y	S	D	Y
C	F	T	U	B	I	G	A	V	X	Z
C	W	E	D	N	E	S	D	A	Y	Y
Y	A	D	F	M	P	R	S	A	Y	Y
A	T	X	A	Q	U	A	E	A	Y	Y
D	N	N	P	G	T	K	U	L	I	F
N	W	P	M	U	N	O	T	Z	E	R
O	A	A	R	D	A	Y	Y	A	C	I
M	W	D	V	U	T	A	B	H	K	D
M	A	D	M	U	T	A	T	A	Y	A
Y	R	S	T	H	U	R	S	D	A	Y

Sunday

Monday

Tuesday

Wednesday

Thursday

Friday

Saturday

Circle the months that are not in the proper order.

JANUARY						
S	M	T	W	T	F	S
		1	2	3	4	5
6	7	8	9	10	11	12
13	14	15	16	17	18	19
20	21	22	23	24	25	26
27	28	29	30	31		

FEBRUARY						
S	M	T	W	T	F	S
					1	2
3	4	5	6	7	8	9
10	11	12	13	14	15	16
17	18	19	20	21	22	23
24	25	26	27	28	29	

MARCH						
S	M	T	W	T	F	S
						1
2	3	4	5	6	7	8
9	10	11	12	13	14	15
16	17	18	19	20	21	22
23	24	25	26	27	28	29
30	31					

APRIL						
S	M	T	W	T	F	S
		1	2	3	4	5
6	7	8	9	10	11	12
13	14	15	16	17	18	19
20	21	22	23	24	25	26
27	28	29	30			

MAY						
S	M	T	W	T	F	S
				1	2	3
4	5	6	7	8	9	10
11	12	13	14	15	16	17
18	19	20	21	22	23	24
25	26	27	28	29	30	31

JUNE						
S	M	T	W	T	F	S
1	2	3	4	5	6	7
8	9	10	11	12	13	14
15	16	17	18	19	20	21
22	23	24	25	26	27	28
29	30					

AUGUST						
S	M	T	W	T	F	S
					1	2
3	4	5	6	7	8	9
10	11	12	13	14	15	16
17	18	19	20	21	22	23
24	25	26	27	28	29	30
31						

JULY						
S	M	T	W	T	F	S
		1	2	3	4	5
6	7	8	9	10	11	12
13	14	15	16	17	18	19
20	21	22	23	24	25	26
27	28	29	30	31		

SEPTEMBER						
S	M	T	W	T	F	S
	1	2	3	4	5	6
7	8	9	10	11	12	13
14	15	16	17	18	19	20
21	22	23	24	25	26	27
28	29	30				

NOVEMBER						
S	M	T	W	T	F	S
						1
2	3	4	5	6	7	8
9	10	11	12	13	14	15
16	17	18	19	20	21	22
23	24	25	26	27	28	29
30						

OCTOBER						
S	M	T	W	T	F	S
			1	2	3	4
5	6	7	8	9	10	11
12	13	14	15	16	17	18
19	20	21	22	23	24	25
26	27	28	29	30	31	

DECEMBER						
S	M	T	W	T	F	S
	1	2	3	4	5	6
7	8	9	10	11	12	13
14	15	16	17	18	19	20
21	22	23	24	25	26	27
28	29	30	31			

Seasons

In Boston, Massachusetts the year has four very different seasons. In summer it is sunny and hot, in winter it snows and is very cold, in spring the flowers bloom and in the fall the trees lose their leaves before winter.

On the calendar below draw a line from the highlighted month with the icon that best represents the weather for that month.

Reading a Calendar
Use what you know about days of the week to complete this exercise.

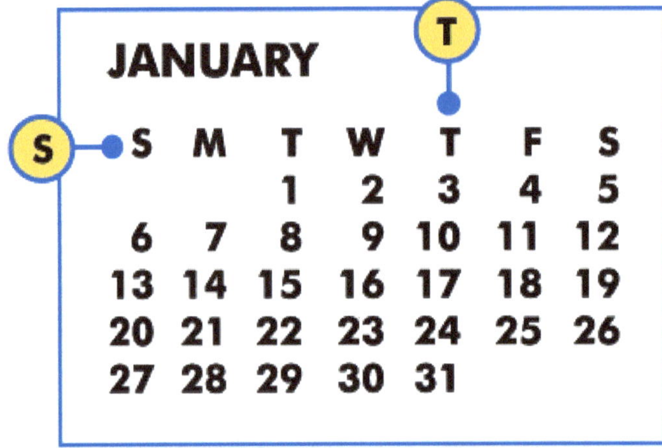

S What day of the week is this?   **T** What about this?

What day of the week is January 11?

What is the date of the third Wednesday?

How many Mondays are there in this January?

This is the calendar for 2036. The last time the calendar was exactly like this was 2008.

2036

JANUARY						
S	M	T	W	T	F	S
		1	2	3	4	5
6	7	8	9	10	11	12
13	14	15	16	17	18	19
20	21	22	23	24	25	26
27	28	29	30	31		

FEBRUARY						
S	M	T	W	T	F	S
					1	2
3	4	5	6	7	8	9
10	11	12	13	14	15	16
17	18	19	20	21	22	23
24	25	26	27	28	29	

MARCH						
S	M	T	W	T	F	S
						1
2	3	4	5	6	7	8
9	10	11	12	13	14	15
16	17	18	19	20	21	22
23	24	25	26	27	28	29
30	31					

APRIL						
S	M	T	W	T	F	S
		1	2	3	4	5
6	7	8	9	10	11	12
13	14	15	16	17	18	19
20	21	22	23	24	25	26
27	28	29	30			

MAY						
S	M	T	W	T	F	S
				1	2	3
4	5	6	7	8	9	10
11	12	13	14	15	16	17
18	19	20	21	22	23	24
25	26	27	28	29	30	31

JUNE						
S	M	T	W	T	F	S
1	2	3	4	5	6	7
8	9	10	11	12	13	14
15	16	17	18	19	20	21
22	23	24	25	26	27	28
29	30					

JULY						
S	M	T	W	T	F	S
		1	2	3	4	5
6	7	8	9	10	11	12
13	14	15	16	17	18	19
20	21	22	23	24	25	26
27	28	29	30	31		

AUGUST						
S	M	T	W	T	F	S
					1	2
3	4	5	6	7	8	9
10	11	12	13	14	15	16
17	18	19	20	21	22	23
24	25	26	27	28	29	30
31						

SEPTEMBER						
S	M	T	W	T	F	S
	1	2	3	4	5	6
7	8	9	10	11	12	13
14	15	16	17	18	19	20
21	22	23	24	25	26	27
28	29	30				

OCTOBER						
S	M	T	W	T	F	S
			1	2	3	4
5	6	7	8	9	10	11
12	13	14	15	16	17	18
19	20	21	22	23	24	25
26	27	28	29	30	31	

NOVEMBER						
S	M	T	W	T	F	S
						1
2	3	4	5	6	7	8
9	10	11	12	13	14	15
16	17	18	19	20	21	22
23	24	25	26	27	28	29
30						

DECEMBER						
S	M	T	W	T	F	S
	1	2	3	4	5	6
7	8	9	10	11	12	13
14	15	16	17	18	19	20
21	22	23	24	25	26	27
28	29	30	31			

What day will your birthday be on in the year 2036?

How old will you be then?

How many?
Circle the answer.

Days in a week	8	7	31	25
Months in a year	11	12	13	15
Weeks in a year	31	12	52	36
Days in a year	365	52	500	100
Days in January	31	32	39	29

Name____

Days, Weeks, Months, Years Quiz

Circle of fill in the correct answer.

 **Which month comes after July?**

 September  August  June  March

 **In which month is Thanksgiving?**

 November  October  June  April

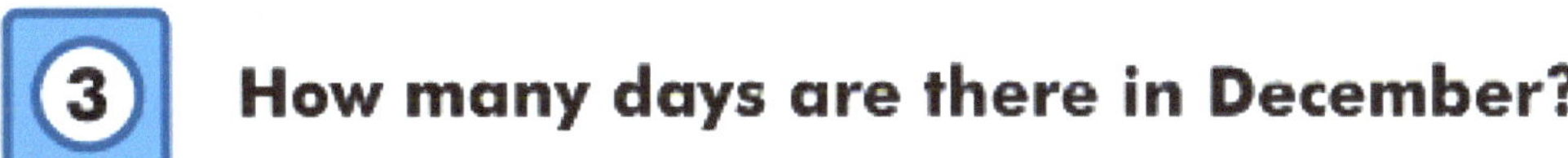

 **How many days are there in 3 weeks?**

Charts & Graphs

Key Vocabulary

tally charts

bar graph

pictograph

survey

data

It's flu season.
Complete this table with both numerals and tally marks.

Absence Record for H. Tubman Elementary

Day	Absences – Tally	Number
Monday	ЖЖ ЖЖ	10
Tuesday	IIII	
Wednesday	ЖЖ ЖЖ ЖЖ I	
Thursday		2
Friday		8

Complete the table.

Can you complete this pictograph?
Look closely at the information that you are given.

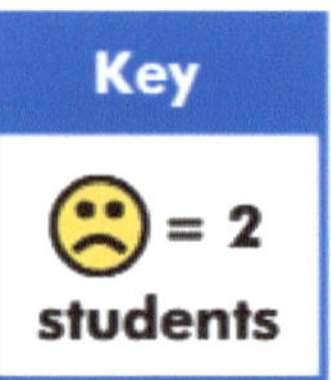

Absence Record for H. Tubman Elementary School

Day	Student Absences
Monday	☹ ☹ ☹ ☹ ☹
Tuesday	☹ ☹
Wednesday	☹ ☹ ☹ ☹ ☹ ☹ ☹ ☹
Thursday	
Friday	

Day	Absences
Monday	10
Tuesday	4
Wednesday	16
Thursday	2
Friday	8

Absence Record for H. Tubman Elementary School

Day	Student Absences	Key
Monday	😦 😦 😦 😦 😦	😦 = 2 students
Tuesday	😦 😦	
Wednesday	😦 😦 😦 😦 😦 😦 😦 😦	
Thursday	😦	
Friday	😦 😦 😦 😦	

On which day of the week were most students absent?

On which day were four students absent?

How many students were absent in total this week?

Which is easier to read, a pictograph or a bar graph? Why?

Absence Record for H. Tubman Elementary School

Day	Student Absences
Monday	
Tuesday	
Wednesday	
Thursday	
Friday	

0 2 4 6 8 10 12 14 16

Absence Record for H. Tubman Elementary School

Day	Student Absences	Key
Monday	☹ ☹ ☹ ☹ ☹	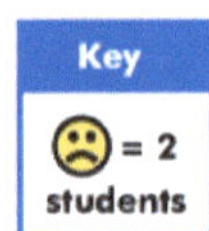 = 2 students
Tuesday	☹ ☹	
Wednesday	☹ ☹ ☹ ☹ ☹ ☹ ☹ ☹	
Thursday	☹	
Friday	☹ ☹ ☹ ☹	

Bar Graph of Student Absences

What day were the most students absent? _______________

What day were the least students absent? _______________

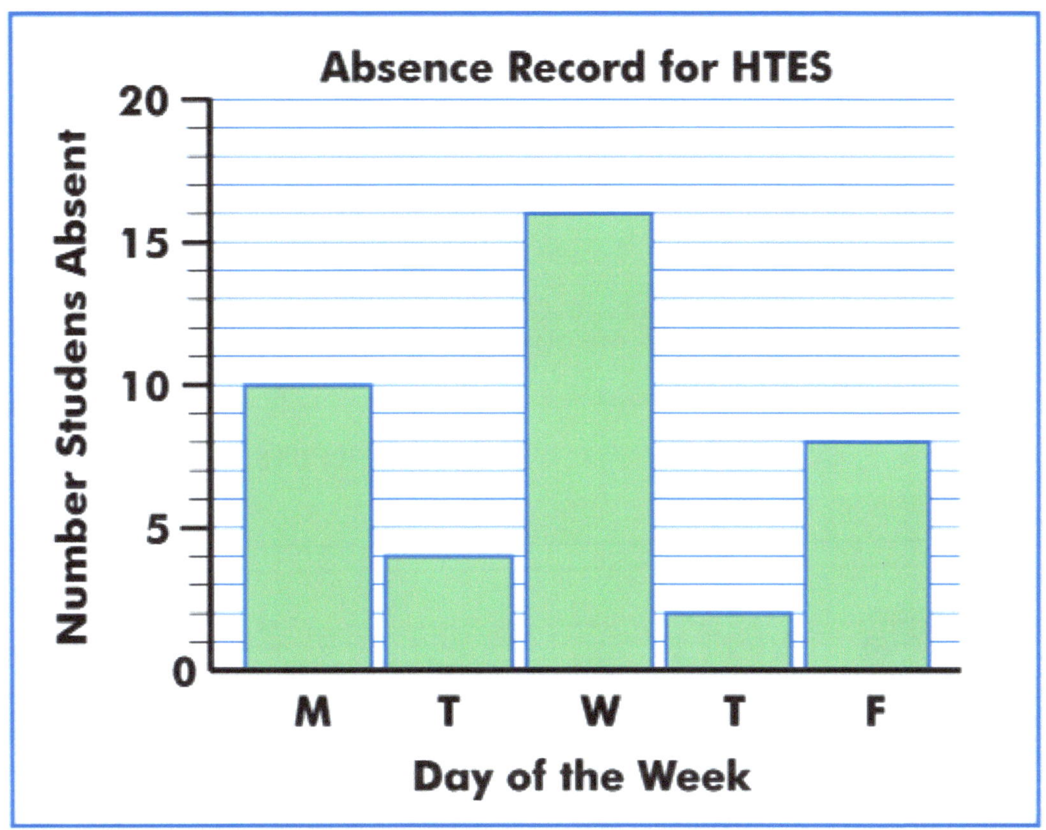

How many students were absent during this week? _______________

Pet Survey

Complete this table of the student pet survey
What surprises you the most about this survey? _______________________________

____

Student Favorite Pet Survey

Pet	Tally (Students)	Number
	ⅢⅡ I	6
	ⅢⅡ III	8
	ⅢⅡ II	
	II	

Complete this graph of the pet survey.

Draw and color the bars.

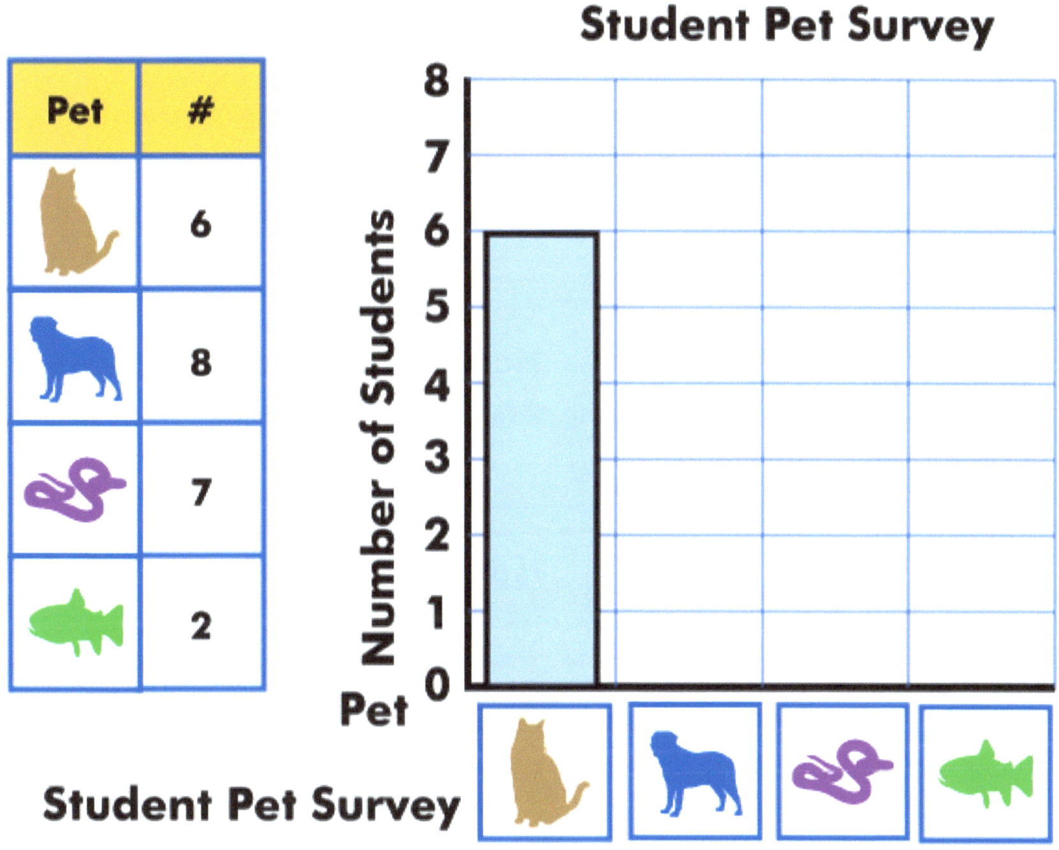

Use the bar graph to answer these questions.

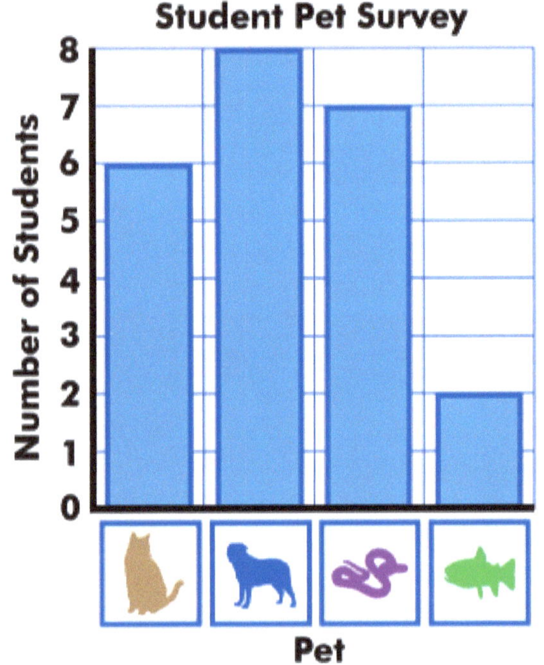

What is the favorite pet?

What is the least favorite pet?

How many more students prefer a dog than prefer a fish?

How many students took part in this survey?*

*Assume each student voted only once

Charts & Graphs Quiz

Circle or fill in the correct answer.

1 **True or false? Owen achieved the highest score?**

2 **How many more points did Mia score than Tori?**

- **A** 2
- **B** 5
- **C** 10
- **D** 20

3 **What was Tony's score?**

4 **What was the combined score of Tony and Tori?**

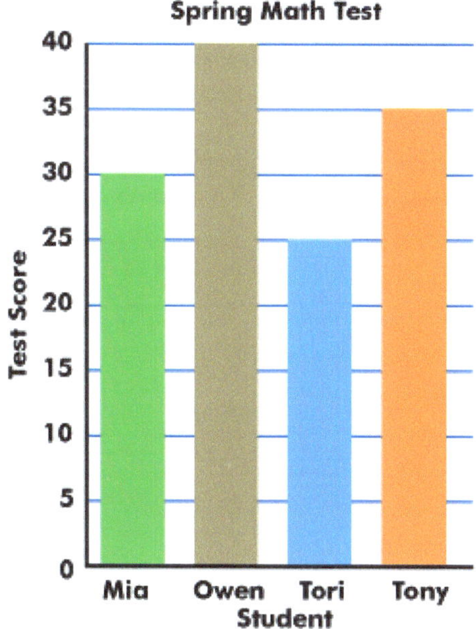

Newburyport, MA 01950

1-800-596-3175

OnBoard Academics employs teachers to make lessons for teachers! We create and publish a wide range of aligned lessons in math, science and ELA for use on most EdTech devices including whiteboard, tablets, computers and pdfs for printing.

All of our lessons are aligned to the common core, the Next Generation Science Standards and all state standards.

If you like our products please visit our website for information on individual lessons, teachers licenses, building licenses, district licenses and subscriptions.

Thank you for using OnBoard Academic products.

www.ingramcontent.com/pod-product-compliance
Lightning Source LLC
Chambersburg PA
CBHW042128030726
47599CB00002B/394

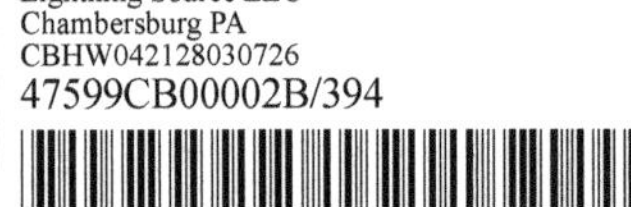